SONORAN DESERT

THE STORY BEHIND THE SCENERY®

by Christopher L. Helms

Director of Development, Arizona-Sonora Desert Museum

The text of this book is based upon material contributed by certain past and present staff members of the Arizona-Sonora Desert Museum, Tucson, Arizona. The expert assistance and willing participation of the following is gratefully acknowledged:

Mark Dimmitt, Ph.D., Curator of Botany
Merritt S. Keasey III, Former Curator of Small Animals
Inge Poglayen, Ph.D., Former Curator of Birds and Mammals
Anna Domitrovic, Assistant Curator of Geology
Richard Felger, Ph.D., Former Staff Ecologist
David Thayer, Curator of Geology

Any endeavor in which the museum plays a part must necessarily bear the imprint of its founders, the late William H. Carr and the late Arthur N. Pack. This publication is therefore a tribute to these remarkable and talented men.

Cover and inside-cover photos by David Muench. Photo this page by Josef Muench.

Book Design by K. C. DenDooven

Fourth Printing, 1991

LC 80-82918. ISBN 0-916122-71-9.

In the midst of the word he was trying to say,
In the midst of his laughter and glee,
He had softly and suddenly vanished away—
For the Snark was a Boojum you see.
—Lewis Carroll, *The Hunting of the Snark*

It seems somehow appropriate that the man who gave the boojum tree its name drew his inspiration from Lewis Carroll, the genius who created the world that is so strange yet so familiar to all of us who in childhood traveled "through the looking glass." For the land in which this master of fantasy took us is no farther from reality and no more familiar than the conception that many people hold of the desert, the place where that strange plant the boojum dwells.

This bizarre creation of nature (it looks like a giant, topsy-turvy carrot and grows in areas that can only be described as "otherworldly") epitomizes in many ways the land that most of us envision when we hear the word *desert.* But more than that, and perhaps somewhat paradoxically, the boojum is the "living proof" that one doesn't have to resort to imagination to find unique and marvelous things. This plant is only one part of an infinitely larger world—a vast desert that lies in America's great Southwest. A place that in reality is always fascinating—and sometimes exceedingly strange. A place that is not at all—or perhaps we should say *much more than*—what we expect it to be.

What, then, do we expect of deserts? Endless, undulating sand dunes occasionally rippled by dry, scorching winds? Gila monsters so vicious that they attack on sight? Blood-red sun beating down with blistering intensity on terrain so hostile that it is utterly devoid of delicacy and beauty? Parched throats crying out for *water,* that precious commodity which is nowhere to be found?

The myths live on, tenaciously. And they are fed by the filmmakers, novelists, and television producers who, less creative than Carroll, find it more advantageous to perpetuate the myths than to create a new fantasy or, far better, to present the incredible reality. And why *should* they work that hard? We are a ready-made audience, so fascinated are we with the scenes that in imagination are just waiting to flash into mind at the slightest mention of "the desert." Just how popular, after all, are the debunkers? It takes a lot of determination to fight a myth, especially one that is held as dear as this one.

But try we must. For in the reality of the desert lies true beauty, a beauty that is known to all too few. And, with Milton, we believe that

Beauty . . . Nature's coin, must not be hoarded,
But must be current, and the good thereof
Consists in mutual and partaken bliss.

PETER KRESAN

Actually, there is a kernel of truth in every myth. And so it is with this one. There *are* vast areas of sand dunes in the desert. A few desert residents *do* bite. The harshness of some desert environments precludes almost *all* life. And water in these particular areas is indeed scarce and precious.

But just imagine for a moment a vastly different scene. It is sunset. The sky is aflame with crimson hues that seem to be more brilliant than any seen before. Against the vibrant sky, magnificent bighorn sheep, standing on an isolated mountain peak, are starkly silhouetted. In the valley below, wildflowers interweave their bright colors in a lush carpet of delicate beauty. Icy streams, teeming with trout, tumble past while deer, unaware of the mountain lion that prowls nearby, partake of the water. Overhead, a hawk catches the last breeze of the day, soaring and wheeling ever higher in graceful, hypnotic motion.

An idealized scene, you might say, and far removed from the desert. But you would be wrong. For not only *is* there such a place, it is also a part of a great region that is dominated by desert—the Sonoran Desert. Within its thousands and thousands of square miles, this desert encompasses such an extensive variety of flora and fauna that botanists and zoologists have found it to be one of the richest areas in the world. Moreover, the topographical characteristics the Sonoran Desert region displays are so diverse that almost any setting imaginable can be found somewhere within its boundaries. For example, it is possible to find here, in this desert region, rolling hills and grasses so tall that one might swear he is standing in the midst of a prairie of long ago. (Indeed, the makers of the motion picture *Oklahoma!* thought the illusion so perfect that they used Sonoran Desert sites for many of the movie's scenes.)

The exotic forms of cholla cacti, silhouetted by a classic western sunset, invoke a deep sense of well-being and a mood of quiet introspection. But the solitude that an evening on the desert seems to promise is illusory. It is only at night, after the blistering heat subsides and darkness provides a protective covering, that most desert creatures pursue their daily activities.

DAVID MUENCH

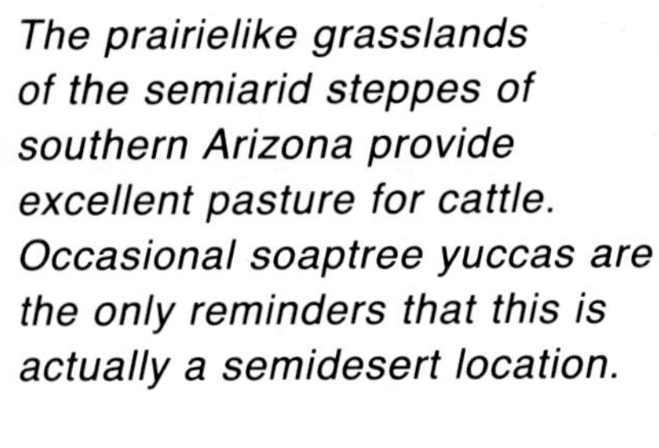

The prairielike grasslands of the semiarid steppes of southern Arizona provide excellent pasture for cattle. Occasional soaptree yuccas are the only reminders that this is actually a semidesert location.

DAVID MUENCH

Thus we have a region, the Sonoran Desert, that offers many contradictions and includes many extremes. Temperatures range from below freezing to well above 100° Fahrenheit. Plants range from the bizarre boojum tree (with which we have already become acquainted) and the saguaro cactus to the familiar ponderosa pine. In some areas, streams support tiny pupfish or three-foot-long razorback suckers, while other

areas are so dry that their inhabitants must resort to some very strange strategies in order to survive. Even stranger things may come out of this desert. It is thought that some of the indigenous plants of the Sonoran Desert may become of life-saving importance in feeding populations of the future. And there is the possibility that one of these plants, the jojoba, will save the now endangered sperm whale from extinction. Liquid wax from the nut of the female jojoba plant is an excellent substitute for sperm whale oil.

It is impossible to describe the animals that live here without also describing the abundant and curious vegetation upon which these animals depend. There is also the land—the foundation to which all life is bound—and how it came to be the way it is. The result is a story of life that is mutually dependent and beautifully interrelated. It is the story of the Sonoran Desert.

There is no one—from casual visitor to dedicated scientist—who won't find something in this desert to fascinate and delight him. It encom-

No, the Pinacate beetle has not stumbled. This desert insect possesses a defense mechanism similar to that of the skunk. When disturbed, this beetle raises its abdomen and secretes a foul-smelling substance.

PETER KRESAN

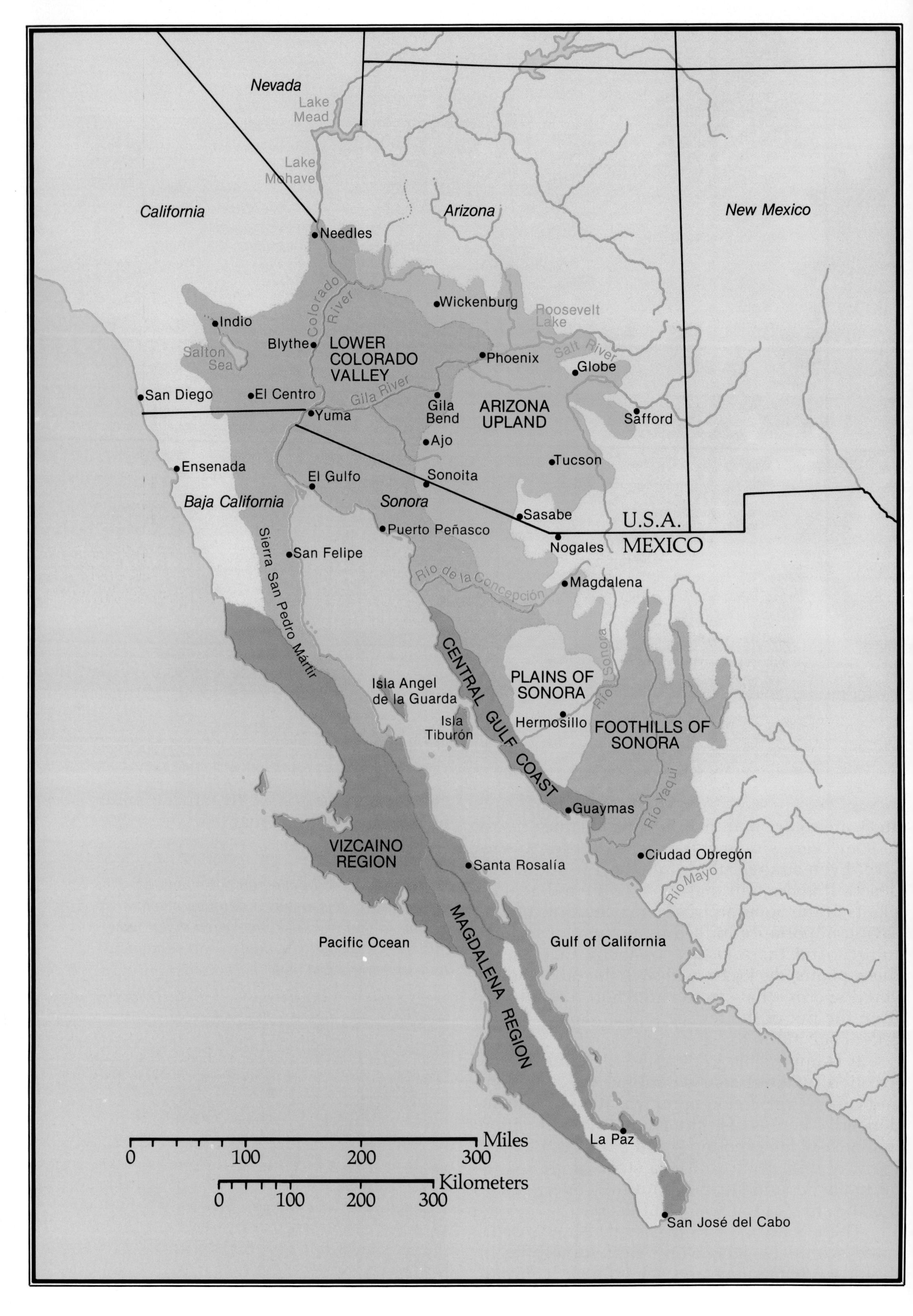
Nevada
Lake Mead
Lake Mohave
California
Arizona
New Mexico
Needles
Colorado River
Wickenburg
Roosevelt Lake
Indio
Blythe
LOWER COLORADO VALLEY
Salton Sea
Phoenix
Salt River
Globe
San Diego
El Centro
Gila River
Yuma
Gila Bend
ARIZONA UPLAND
Safford
Ajo
Tucson
Ensenada
El Gulfo
Sonoita
Baja California
Sonora
Sasabe
U.S.A.
MEXICO
Puerto Peñasco
Nogales
San Felipe
Sierra San Pedro Mártir
Rio de la Concepción
Magdalena
CENTRAL GULF COAST
Rio Sonora
PLAINS OF SONORA
Isla Angel de la Guarda
Isla Tiburón
Hermosillo
FOOTHILLS OF SONORA
Rio Yaqui
Guaymas
VIZCAINO REGION
Santa Rosalía
Ciudad Obregón
Rio Mayo
MAGDALENA REGION
Pacific Ocean
Gulf of California
La Paz
San José del Cabo
Miles
0 100 200 300
Kilometers
0 100 200 300

passes many ecosystems, each of which is finely tuned to accommodate some of the most interesting and exciting life on this planet.

The story that follows, then, is a true one, although it contains elements that are stranger than fiction. Indeed, as we go along, the tale will become—if we may borrow again from the delightful vocabulary of Carroll—"curiouser and curiouser"!

A Living Desert

The Sonoran Desert covers approximately 120,000 square miles, a huge area that includes southwestern Arizona, southeastern California, most of the peninsula of Baja California (including the islands in the Gulf of California), and the state of Sonora, Mexico. It is one of the four deserts on the North American continent: the Sonoran, the Great Basin, the Mojave, and the Chihuahuan.

Creosote is the most prevalent plant in the Sonoran Desert.

Deserts—areas characterized by scant water and sparse vegetation—occupy nearly a fifth of the earth's surface. For mapping purposes, the four major North American deserts have been delineated by the vegetation that is unique to each, although many of the same species are represented in all four. The Sonoran Desert is the most species-rich, and it encompasses an unusually broad diversity of environments.

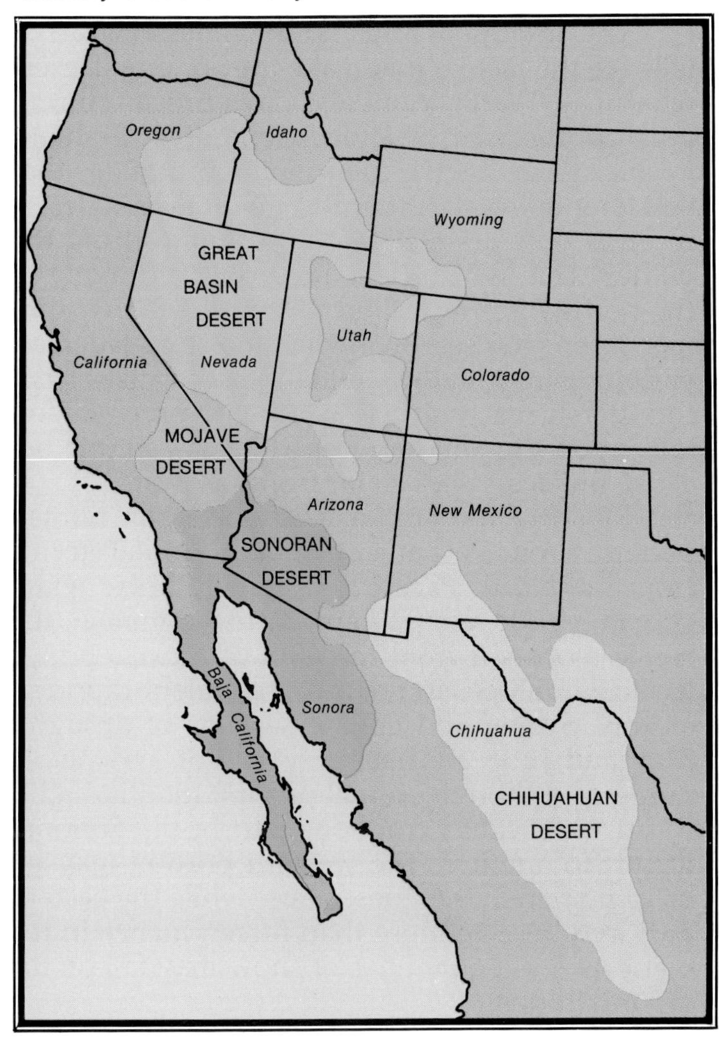

The Sonoran Desert, for the most part, is a low, hot desert. Its lowest elevation (the area that surrounds the lower Colorado River) rivals Death Valley as the hottest and driest place in North America. Summer highs frequently exceed 120° F., and surface temperatures often approach 200°! Under a cloudless sky, when the humidity is less than ten percent, solar radiation is so intense that it extracts the life-giving water from an exposed plant—water that is all the more precious because it usually cannot be replaced through the plant's roots as they spread in searching tentacles through the parched mineral soil. Annual rainfall in this Colorado subregion averages less than three inches, and there have been periods of more than two years in which there was no rain at all. Even so, plants *do* exist—lots of them—even in the most desolate portions of this desert. They exist because, like desert animals, they have been able to adapt to the conditions that this harsh environment imposes.

The creosote bush, for example, is a classic desert plant. It exhibits a typical desert adaptation in the development of leaves that are small enough to check the loss of water during periods of intense heat and dry air. This rather nondescript shrub is the most prevalent plant in the Sonoran Desert, probably because it is the most drought-adapted perennial here. It can drop its leaves to reduce water loss and, during the worst droughts, it even dies back to the ground. Then, when it finally rains, the plant revives and flourishes—like the mythological phoenix. Despite the small stature of creosote plants (they rarely attain a height or breadth of over five feet), they often live to be very old. Some surviving clonal material is thought to be twice as old as the bristlecone pine, which is the oldest of living things, some specimens still surviving at ages of four to five thousand years.

The common denominator of *all* deserts is a lack of moisture—a factor that is modified by seasonal fluctuations, duration and intensity of rains, rate of evaporation, nature of the soil, and availability of water. (The latter condition is exemplified by frozen arctic regions of Europe and North America. There is water, but it is usually frozen and thus is of little use in supporting plants.) But deserts throughout the world have other characteristics; these differences generally place them into one of the two main categories: *horse-latitude* and *rain-shadow.*

Horse-latitude deserts are found on the western side of every continent at 30° north and south latitudes, where a year-round trend of descending—and therefore warm and dry—air prevails.

Rain-shadow deserts can occur anywhere on the lee sides of mountain ranges. Since water has been drained from the air while passing over the mountains, little moisture is left for the other side. The remote interior deserts can be said to be a kind of rain-shadow desert in that their great distances of flat land have the same drying effect as a mountain range.

The Sonoran Desert is a combination of both types—horse-latitude at 30° and largely rain-shadow in the northern limits. It is quite lush when compared to other deserts of the world. Unlike the other three North American deserts, the Sonoran Desert straddles the frost line; thus most of it is frost-free, or nearly so. Its vegetation is dominated by two life forms—legume trees and columnar cacti, but it nevertheless encompasses a rich spectrum of some 2,500 species of plants. Such diversity, needless to say, enlivens and greatly enhances the Sonoran Desert landscape.

The contrasts contained within the Sonoran Desert region are incredible. Its "typical" desert areas, which harbor nearly 2,500 species of plants, are expected, but the region also includes mountain ranges that may be densely forested and are sometimes covered with snow.

PAUL R JOHNSON

A rainbow blesses the Sonoran Desert after rain has ministered to the needs of its plants, soil, and animals.

The tremendous variety of plant and animal life in the Sonoran region is what makes it all so difficult for many of us—burdened as we are by our myth-derived conceptions—to accept the Sonoran as *desert.* The Sonoran region has large pockets of arid and semi-arid "typical" desert terrain. It also contains mountain ranges whose upper slopes are densely forested with pine and fir. And it even includes offshore islands, in the Gulf of California, which are visited by gulls, terns, and other sea birds.

The amount of rain and the times at which it falls are the factors that make the Sonoran Desert what it is. Warm, cloudless days prevail during much of the year. Shielded from large, frequent Pacific storms by the high mountain ranges that lie along its western border, the Sonoran Desert receives only a small amount of rain during the winter. That which does occur falls primarily in the northwestern portion. Even during the summer "monsoon season," when moisture-laden air sweeping in from the Gulf of Mexico causes large thunderstorms, only a few inches of precipitation falls upon the Sonoran Desert itself—simply because the desert lies so far from the gulf.

Much of the area has a two-season rainfall pattern. From December to March, frontal storms from the North Pacific occasionally bring widespread, gentle rains. From July to September, the tropic monsoon from the Gulf of Mexico brings localized deluges in the form of violent thunderstorms. Spring and fall are seasons of drought. May and June are usually very hot and nearly rainless. Although annual precipitation is slight, it can vary widely from one part of the Sonoran Desert to another. The western portion may receive no rain in the driest years, while the eastern part may receive more than fifteen inches in the wettest. Great quantities of snow may fall on the mountaintops.

RICK McINTYRE

The Gila monster is a docile reptile rarely seen in the wild. Protected by law, it is one of only two venomous lizards in the world; the other is the Mexican beaded lizard.

The small packrat builds a nest that is nothing short of stupendous.

The other North American deserts are colder, with one annual period of rain. The Mojave Desert has a winter rainy season, and most of it experiences severe frosts. Vegetation in the Mojave is composed of a sparse cover of low shrubs; few trees and succulents exist there. The Chihuahuan Desert also lies at a fairly high elevation, where a severe winter and a late-summer rainy season is the norm. The vegetation of the Chihuahuan consists largely of low shrubs and leaf succulents, but there are trees that live along the drainage courses and there are numerous small cacti and other succulents. The Great Basin Desert is very high (above 4,000 feet), with very cold winters. Although precipitation falls over a long season, the summer growing season is short. Vegetation consists of a few species of low, small-leafed shrubs; there are almost no trees or succulents.

The boundaries of the North American deserts cannot be exactly defined, of course. They are loosely determined by both the plants and animals of each region, although generally the important differences are botanical. The matter of where each species of plant and animal will live depends upon temperature, elevation, rainfall, soil type, and terrain. Many species of birds, mammals, reptiles, and plants can be found living in all four deserts, but each desert has some species that it can claim as primarily its own. Essentially, nowhere but in the Sonoran Desert, for example, will one find the reticulated Gila monster,one of only two venomous lizards in the world.

This reptile, by the way, has a fierce reputation but is actually not the treacherous devil it's made out to be. It will attack only when cornered and provoked. One of the myths that constantly maligns this inoffensive creature is that when it bites it doesn't let go until the sun goes down. Not true, of course. The myth probably stems from the manner in which the Gila monster must inject its venom. Unlike the rattlesnake, whose hollow fangs can pierce and inject its venom instantly, the fangless Gila monster must bite, latch on to its victim with its many short teeth, and grind its jaws so that the venom, mixed with saliva, can enter the wound and do its work effectively. Not a very nice prospect, to be sure, but one you are not likely to encounter as long as you only look, not try to touch, if you spot a Gila monster in the wild. Most animals avoid contact with humans, but they will protect themselves if threatened.

Of Caves and Packrats

Parts of the Earth Sciences Center of the renowned Arizona-Sonora Desert Museum—which lies fourteen miles west of Tucson, Arizona, and contains exhibits that are considered to be among the most innovative concepts of natural-history interpretation in existence today—are man-made "wet" and "dry" limestone caves. Built underground, the caves are complete with stalactites and stalagmites fashioned so artfully that visitors often doggedly insist that they are *real.*

Within these caves people can experience the thrill of "discovering" delicate or massive decorations hundreds of thousands of years old, a thrill usually known only to the dedicated spelunker. Moreover, the caves have proven to be highly effective devices by which people can gain an accurate understanding of the fascinating nature of the Sonoran Desert region, including its geology, plants, and animals.

Caves in the Sonoran Desert are by no means the most dominant of the region's physiographic features. However, by finding out how they were formed and by studying the remains of past in-

TOM MYERS

spines until the joint falls off, then pulls out the spines stuck in its skin. Many other animals would panic in such a situation.

The packrat uses anything it can possibly carry to build and embellish its nest. (This was of course how it got its name.) Researchers have even discovered watches, rings, and other strange artifacts in packrat nests. Losses of these items must have greatly perplexed their original owners, and one can't help wondering how many innocent people might have had to take the blame for the thefts of these clever little kleptomaniacs!

It takes a considerable amount of time for a packrat to complete its nest (if indeed such a nest can ever be considered "completed"). In the effort expended and the size (proportionally speaking) of the result, the project is comparable to the building of an entire family-sized house by one

habitants or intruders, we can learn a great deal about the typical plants and animals that lived thousands of years ago—not only in the caves but also in the entire area. This includes humans, who have been present in this region and have used the caves for at least 6,000 years. And it includes the packrat.

The packrat very frequently inhabits caves throughout the Sonoran Desert. Referred to affectionately by some museum professionals as "our first curators," these desert dwellers have taught us valuable lessons in the methods by which animals have adapted to the desert. They and the other occupants of their unusual nests provide excellent examples of how interrelationships allow animals to exist in harsh environments.

The packrat, a nocturnal animal like many other desert creatures, spends most of its time engrossed in two activities: eating and nest-building. A packrat usually looks for a convenient cholla or prickly-pear cactus to use as a foundation for its huge nest, but sometimes it seeks out a secluded spot under a tree or in a cave, abandoned shack, or basement. These rodents sometimes build nests as large as six feet across and three feet high!

Living cacti are preferred as nesting sites because the spines provide protection by discouraging large predators. Also, the packrat can use the plant itself as a source of food. No need to worry that a packrat might stick itself with the spines. This animal can easily handle the cholla joints (segmented parts of cholla stems). It simply removes an embedded joint by biting through the

An avid cave explorer investigates the mysteries of a formation that was created by water millions of years ago in the Sonoran Desert region.

GARY LADD

person. The packrat is a solitary creature, who almost always works alone.

But the packrat is practical, too, and is not opposed to occupying an abandoned nest if it can find one. Scientists have discovered packrat nests that contain artifacts as old as 10,000 to 14,000 years, and the evidence indicates that the nests have been in continual use all that time. The "urine cement" with which the packrat binds his nest is capable of preserving a packrat *midden* (junk pile) for centuries.

Picking a midden apart can be rather like looking back in time, an experience that is not only fascinating but also educational. Analyses of the contents of middens often reveal important facts about plants and animals that existed in a particular area thousands of years ago. Middens are especially valuable in this respect because of the fact that packrats tend to stay within only a short radius of home. In its Earth Sciences Center, the Arizona-Sonora Desert Museum exhibits a midden that when found contained seeds and some prehistoric mammal and bird bones nearly 12,000 years old!

The spaciousness of the packrat's nest invites many other creatures to avail themselves of its advantages. Some use the structures for protection; some use them just to cool off in when the heat is at its most oppressive (the nests are naturally air-conditioned); and some move in permanently. The blood-sucking conenose, or kissing bug, for example, makes its home in a packrat nest, feeding off the rodent and venturing out only now and then to supplement its diet with the blood of some other creature. The black widow and brown recluse spiders also often take up lodgings in this desert "apartment," as do lizards and several species of snakes and scorpions. Thus a remarkably integrated and odd assortment of tenants is indebted to the packrat: builder, host, and landlord of the nest.

How Did the Desert Come to Be?

There are five significant chapters in the geologic story of the Sonoran Desert.

Chapter One is about "basement rocks" and plate tectonics. At the very bottoms of our stark volcanic mountains, at the utmost foundations of the desert basins, and in occasional outcrops where they have been violently thrust to the surface, we find the earliest formed rocks of the region. These are baked and crushed basement rocks of the ancient core of our desert: the Pinal Schist, one thousand seven hundred million years old.

The world is thought to be about four and one-half billion years old, so our planet was already ancient when these foundation rocks of the desert formed. And the story of *how* they formed is the story of the origin of Earth's crust itself. According to the accepted theory of plate tectonics, the earth is divided into a small number of crustal "plates" that float like lily pads on the semimolten rocks of our planet's interior. Unlike lily pads, the plates are in constant motion, driven by internal fires and traveling at the speed that fingernails grow. The continents, like toads on the lily pads, are rafted about as the plates separate and recombine.

When two plates collide they create a great deal of friction, as you can well imagine. This

friction melts rocks at the edge of continents, forming volcanic and underground blobs of molten magma. These igneous rocks are continually added to the existing continents, increasing the sizes of these continents through time. By this and other processes our North American continent grew to its present size. The basement rocks of the Sonoran Desert region were "plastered on" about 1,700,000,000 years ago. For a billion years thereafter, this basement was subjected to erosion and upheavals until it came to be a flat coastal plain at sea level.

Chapter Two encompasses 350 million years when various seas flooded our region, deposited their loads of sand and lime, and then retreated.

Chapter Three is the Age of Dinosaurs—a time when the desert was subjected to an even more intense buildup of volcanic activity, reaching its crescendo at the end of this era, sixty-five million years ago. Many of our mightiest mountains formed during this violent stage, including the Santa Catalinas and the Santa Ritas. Most of the region's immense copper deposits were also formed at this time.

Chapter Four is the Age of Mammals. In the middle of this era, about twenty to thirty million years ago, another round of violent volcanism occurred. No sooner had the volcanic dust settled than great faults, or breaks in the earth's crust, ripped the region into long north–south segments. The desert landscape was being stretched by movements of the great Pacific and North American tectonic plates. Alternate fault segments rose or fell during the interval from fifteen to eighteen million years ago, giving shape to our present-day basins and mountain ranges of the Sonoran Desert area.

Chapter Five is the story of the Ice Ages and the arrival of humans on the desert scene. This is technically a short, recent paragraph of Chapter Four, the Age of Mammals, but we give it special emphasis here because it is so recent that we have greater knowledge of its history, and because its history concerns human beings.

The Ice Ages gripped our desert region not with scouring glaciers but with changes in climate. During the times that one third of North America was ice-covered, the Sonoran Desert region was relatively free of ice but was cooler and wetter than it is now. Oak, pine, and juniper dotted slopes where saguaros and acacias grow today. Mammoths, horses, and giant camels roamed the basins along water courses that are now mostly dry.

Toward the end of this period, about 11,000 years ago, humans arrived. First were the big-game hunters, then the archaic hunter-gatherers, then the pottery makers: the Hohokam, or "those who have vanished." They were here almost until the Spaniards arrived in the New World, and they may be here even today. There is speculation that the modern Tohono O'otam Indians (formerly known as the Papago Indians) descended from the Hohokam.

Today as we examine the natural history of the Sonoran Desert area we find that the modern inhabitants—plants, animals and people—are a product of and dependent upon the 1,700,000,000-year-old geologic foundation beneath our feet.

The Gran Desierto *of Sonora, Mexico, the largest and driest sand sea in North America, is probably representative of what most people think of as "typical" desert. The sand depth of these windswept star dunes varies from a few feet to hundreds of feet.*

PETER KRESAN

t Laguna Salada, ortheast Baja alifornia, the line at e base of the ountains clearly efines a sharp fault at indicates recent bout 1970) arthquake activity. his fault is haracteristic of the eologic dynamics the Sonoran esert region.

Geographically, the Sonoran Desert is always changing. These cinder cones in the Pinacate volcanic field are the result of eruptions that occurred within the last ten thousand years or so. The Pinacate field is dormant but potentially active. Geologists are fascinated by the fresh texture of the lava flows, which look exactly like Hawaiian lava flows from eruptions that took place only yesterday!

PETER KRESAN

Volcanism

Imagine a volcano that is about to erupt. The ground has been trembling for some time now, and occasionally the entire landscape shudders uncontrollably. A faint rumbling can be heard. There is a pervasive tension in the air, an uneasiness that is sensed by all the creatures that live there. Almost imperceptibly the ground swells as underground a melted blob of the earth's crust, *magma,* pushes upward, oozing like toothpaste through layers of rock on its way to the surface.

The surrounding rocks are broken, altered, and decomposed by the rising magma, which—having originally been part of the upper crust of the earth—is very rich in quartz and therefore very viscous. Within this hot (1,000° Centigrade), thick fluid, water vapor and carbon dioxide are dissolved. The pressure within the magma chamber builds.

The rumble is now an ominous roar. The swelling of the ground is unmistakeable. Fissures form, and gases hiss as they escape through the cracks. Suddenly the ground is ripped asunder. Molten rock spews forth in a violent eruption and is transformed into a terrifying, eerily beautiful avalanche that rapidly swallows everything in its path. On the heels of the explosion, a spectacular

PETER KRESAN

Weathering and erosion are major geologic processes that never cease to change and shape the Sonoran Desert landscape. Baboquivari Peak, a distinctive landmark to the south of the museum, looks like a volcanic neck. But the resemblance is deceptive; the peak is composed of granitic rock and was sculpted by weathering and erosion.

fountain of hot (800° C.) ash and dust shoots into the air, settling quickly into an incandescent cloud that rolls across the countryside at a speed of nearly a hundred kilometers an hour. Within minutes, the bulky, glowing magma (now called *lava*) has enveloped the immediate terrain in its fiery embrace. As it continues downward, the glowing ash cloud condenses into rock fragments. These rock fragments, still hot and soft, stick together and become compacted. The result, after cooling, is a hard, volcanic rock known as *welded tuff*. A new landscape has been formed.

The catastrophic event we have just imagined is probably very similar to that which actually happened in the Tucson Mountains about seventy million years ago, and again about twenty million years ago. In fact, as one drives over Gates Pass to the Arizona-Sonora Desert Museum, one travels across the remnants of an ancient volcano. The Tucson Mountains themselves are volcanic because the rocks of which they are made were laid down during many explosive volcanic eruptions. There is no thermal activity evident in these mountains today.

The most spectacular evidences of volcanic activity in the Sonoran Desert are revealed today in the *Gran Desierto* of Mexico, at the head of the Gulf of California, immediately south of the Arizona border. The ancient lava flows of this volcanic field—the *Pinacate*—and accompanying sand dunes occupy about 5,000 square miles, some of the most desolate land that can be found in the Sonoran Desert.

Few animals and plants are able to cope with the rigors of the Pinacate. One wonders how *any* living thing can exist in this bleak environment. But even here, in these spectacular badlands, the Sonoran Desert takes care of its own. Lizards, snakes, rodents, birds, insects, and—on the peaks—bighorn sheep make the Pinacate their home, sharing the landscape with hardy creosotes, mesquites, cacti, and other typical Sonoran Desert vegetation.

The most recent volcanic activity that rocked this area occurred only a few thousand years ago, but as we stand in the eerie stillness of the Pinacate, we are struck by thoughts of the magnitude of the forces that were once unleashed here. The scene before us is the result of an awesome example of nature's might. It represents vast destruction—and monumental creation. One gazes out upon dead craters and lava flows that stretch as far as the eye can see. What a display of pyrotechnics that landscape must have produced.

Plate tectonics, volcanism, erosion, and deposition: powerful forces that are still slowly and continously—and sometime violently and abruptly—reshaping the topography of the planet.

The spectacular drive from Tucson through Gate's Pass on the way to the Arizona-Sonora Desert Museum takes visitors through the remnants of an ancient volcano.

Plants, Soil, and Rain

The geological dynamics in the Sonoran Desert are responsible for the range of soils that exists here. It is especially important to understand the nature of these soils, since the plants they support are the criteria by which the geographical boundaries of the desert are defined. Roger Dunbier, in his book *The Sonoran Desert,* explains it this way:

> *The first ten feet of the earth's surface is the creator and in part the creation of its vegetation. If the surface is composed of solid rock or drifting sand it will probably be unable to support a plant cover and will not usually be termed soil. . . . Some areas of the Sonoran Desert are devoid of soil by this definition. . . . In the study of the natural environment, soil and vegetation cannot be separated, nor can they be examined without constant reference to the climatic conditions under which both are in constant formation. It is also impossible, particularly in a desert milieu, to separate soil from the material out of which it was formed and upon which it lies.*

And so the topography in effect creates the plants, and both are in turn affected by climatic factors, particularly rainfall. In deserts, not only is there a chronic deficiency of water, but the rainfall is extremely variable from year to year. In the Sonoran Desert, one or both of its annual "rainy seasons" frequently fail to bring rain in the quantities that make it biologically effective, but occasionally torrential downpours occur that within minutes inundate the land with great rivers of water. A two-year drought may be broken by a single storm that in one hour may drop more than the annual average. Incidentally, the short-term rainfall record for the United States is held by a location that lies just a few miles outside the Sonoran Desert. In 1891 Campo, California, received eleven inches of rain in eighty minutes.

Survival Strategies

There are several "subdivisions" of the Sonoran Desert, each of which has its own characteristic vegetation and topography. The plants and animals that live in these differing areas have evolved numerous and ingenious survival techniques that enable them to cope year after year with the rigors of their particular habitats.

One of these subdivisions, the Arizona Upland, most of which lies in south-central Arizona, is particularly interesting. It contains many more mountain ranges than the other areas of the So-

Jumping cholla cacti, acacia, and wild asters await the nurturing rain of an impending desert storm in the Santa Rita Mountains of southern Arizona.

noran Desert. The vegetation and animal life are abundant.

That common desert plant the creosote is very abundant in the Arizona Upland, but it is the several species of legume trees and cacti that are the most conspicuous plants here.

The most common of the legume trees is the *paloverde* ("green stick"), named for its green bark, which carries out the process of photosynthesis even when the tree is leafless. The two varieties represented here are the blue paloverde and the foothill paloverde. The other two most prevalent legume trees are the ironwood and the velvet mesquite. All of these trees bear prodigious quantities of highly nutritious seeds.

Arizona-Sonora Desert Museum researchers have long been engaged in the intriguing work of investigating the potential of desert plants as commercial crops for the twenty-first century. Some see a certain irony in this research, since prehistoric inhabitants of the Sonoran Desert are known to have utilized some 450 species of these plants as sources for food, and many of these species were major staples as well.

The research may prove to be beneficial, if not crucial, in the feeding of populations of the future. The most important food crops of the world now number only several dozen, and seven of these (wheat, rice, maize or corn, barley, soybeans, common beans, and potatoes) feed most of the earth's people. But worldwide there are probably thousands of plants that have the potential to become major new "crops" for modern agriculture. It is the aim of researchers to use these diverse indigenous crops to develop agricultural products specifically adapted to local environmental conditions. In other words, we may find it possible to develop crops that will fit a particular environment instead of, as is now the case, modifying the environment to suit the crop.

Among the Sonoran Desert plants that appear to have the most potential as new crops are the very trees that are so numerous here: the paloverde, ironwood, and mesquite. From early prehistoric times, native peoples in the Sonoran Desert have used the mesquite tree to provide or fashion food, fuel, shelter, weapons, tools, fiber, and medicine. The mesquite possesses many features that make it a prime candidate as a food crop. The pods ripen nearly simultaneously, making harvesting easier. The pods are large, do not split at maturity, and fall when fully ripe. Flour produced from mesquite pods is edible without having to be cooked. (The process requires muscle power but no fuel.) The ironwood and paloverde produce seeds equal in nutritive value to the mesquite's.

These plants are but a few examples of the hundreds of plants indigenous to the Sonoran Desert that may become world staples of the future. Research in this field is only in its infancy, and none of these plants can now be considered competitive with modern crops. Nevertheless, the immense possibilities of the "new crop agriculture" make it a fertile field of study; perhaps it is a timely one as well.

DAVID MUENCH

Colorful cacti and bright poppies enliven the desert below Tonto Ruins National Monument. Look but don't touch the "furry" teddy-bear cholla, segmented cacti whose spines easily impale clothing or flesh.

DAVID MUENCH

The stately saguaro, with its massive presence and showy blossoms, is the undisputed symbol of the Sonoran Desert. No two saguaros look alike. Many grow straight, but some assume a more unusual appearance, such as that of this individual, whose twisted but elegantly graceful arms transform the plant into an extraordinary work of art. Saguaros have a very long life span. This particular plant, which grows in Saguaro National Monument, might well have begun life during the American Revolution; if so, it probably began to produce arms about the time of the Civil War.

The Saguaro—Symbol of the Desert

The magnificent *saguaro* (suh-WAH-row) is the dominant cactus in the Arizona Upland. It is also an oft-used symbol of the Arizona desert. Nary an epic Hollywood western has been made that does not feature the saguaro cactus. This strange sentinel of the Sonoran Desert, which sometimes stands as tall as fifty feet, grows in "forests," embracing other living things in what appears to be an almost parental relationship. One might even say that the saguaro and the many plants and animals that live in harmony with it enjoy a "family" relationship. This miraculous plant is the consummate example of the interrelationship that exists among the plants, birds, insects, reptiles, and mammals (including man) of the natural world.

The mature saguaro is a mammoth; it is about

thirty-five feet tall and weighs several tons. But for all its gigantic stature, the saguaro has as humble and inauspicious a beginning as any plant, germinating as it does from a tiny seed nourished by summer rains. It faces a multitude of hazards in its first decade: summer heat, winter frosts, droughts, floods, and hungry or clumsy animals succeed in making the existence of the young saguaro precarious at best. The only saguaros to survive are those few that start life in the shelter of either large rocks or "nurse plants," which offer some protection from animals and environmental extremes. After about twenty-five years, the surviving plants have grown to about a foot tall and are somewhat less vulnerable, although rabbits and packrats may still damage them by girdling them as they feed.

The saguaro blossom, official flower of the state of Arizona (one of only two states in which the saguaro grows), takes many, many years to appear. In fact, saguaros don't produce their first flowers until they are seven or eight feet tall and are fifty to sixty years old. Arizona achieved statehood in 1912, which means that a saguaro blooming for the first time this year is very likely older than, or *as* old as, the state itself.

After the saguaro has achieved this height, the growth rate slows, and it may take four more years for it to gain another foot. At the age of seventy-five or a hundred years, it may have achieved a height of twelve to twenty feet, at which point arms begin to appear. The buds usually occur at the thickest part of the cactus, seven or eight feet above ground, at the places where the flowers first began to develop. The newly sprung arms begin to produce flowers within two or three years.

Once they have produced arms, no two saguaros look alike. One plant may grow straight and erect, another may be bent and twisted. The position and number of arms create innumerable variations from plant to plant. Some saguaros have only two or three arms; others have ten, twenty, or as many as fifty. Now and then, smaller arms branch out from the first arms, especially if the original arm tip has been damaged.

Occasionally a saguaro, rather than rising in a straight column, grows into a *cristate* (crested) saguaro, a shape that seems wildly outlandish even for this environment. In a cristate, the tip of the growing stem spreads outward in a large fan that may be several times the width of the main trunk. No one can really explain why this happens. Cristates are abnormal but not diseased; they are simply the weirdest form taken by a very weird plant. Cristate saguaros are not exceedingly common, but at least one specimen can usually be seen in any large stand of saguaros.

We have mentioned survival strategies, and

JODY FORSTER

The cristate saguaro, with fanlike top, is abnormal but not diseased.

The white-winged dove, a small desert pigeon, is a major propagator of the saguaro cactus. While feeding upon the nectar in the blossoms, this bird pollinates the plant. When the saguaro bears fruit, the dove eats and then excretes the seeds, thereby distributing them throughout the range of the saguaro.

DIANE ENSIGN-CAUGHEY

The spines of a saguaro cactus are an excellent example of plant adaptation to a severe environment. Although the spines serve to protect the saguaro from live marauders, their most important function is to shade the plant from the blistering heat of the sun. The spines also act as a barrier against desiccating desert winds.

the saguaro exhibits several fascinating ones. Like many other Sonoran Desert plants, saguaros are *succulents,* plants that have the capability of storing water within their leaves, stems, or roots. (All cacti are succulents, but not all succulents are cacti.) Succulents have extensive shallow root systems that quickly absorb large quantities of water—even from very light rainfalls—and store it for use through the dry months. A less visible but equally interesting adaptation of the saguaro is the waterproof, waxy substance the plant secretes on its surface. Water loss can only take place, therefore, through the *stomates,* tiny pores through which photosynthetic gas exchange takes place.

The spines of this Sonoran giant help it to adapt to its unyielding environment in a couple of ways. To deal with the difficult problem of storing water in a land where it is scarce and at the same time keeping this supply protected from thirsty, hungry animals, the saguaro has developed a spiny armor. This armor serves only to deter its enemies, however, and even the spiniest cacti are not completely immune to attack by woodrats, rabbits, big-horn sheep, javelina, and other creatures who can and do eat cacti when they become thirsty or hungry enough. Not as vulnerable to cactus spines are several insects, including cactus beetles, that specialize in diets of cacti.

More important to the survival of the saguaro is the fact that these spines shade the plant from the relentless sun and insulate it against freezing. This may seem unlikely at first thought, but if one examines a saguaro on a sunny day—particularly a small plant—one will see that the dense groupings of spines do indeed cast a great amount of shade upon the outer surface of the plant. The spines also create a barrier against desiccating desert winds.

The water-storage capacity of the saguaro is incredibly large; the plant may actually increase its weight many times in periods of "plentiful" water. To help support the additional weight, the plant has an interior circular column of thirteen to twenty wood ribs which runs up the main stem of the plant and branches into each arm. Saguaro cactus ribs, collected from dead plants, have been used for centuries by the inhabitants of the Sonoran Desert for tools and building materials.

But man is only one of many animals who have learned to use this remarkable plant to their own advantage.

A Desert "Condominium"

Because of its massiveness, the interior of the saguaro maintains a fairly stable temperature. In the daytime this temperature is usually several degrees cooler than that of the air outside. On cool nights, however, the interior of the plant is warmer. This feature makes the saguaro an ideal desert home for any number and species of birds.

The initial foundation for this desert "condominium" is laid by local woodpeckers—Gila woodpeckers and gilded flickers—in a large arm or in the main body of the plant, a procedure that does not harm a healthy saguaro. The Gila woodpecker is a fussy nest-builder. It may build two or three nests in one season (a two- or three-month period) before it is satisfied with the product. All the better for other potential residents! Once a cavity is created and available, screech and elf owls, purple martins, and Wied's crested flycatchers move right in. The cactus wren, Arizona's official state bird, is not primarily a cavity-nester; therefore, it usually does not, as many people believe, nest in the saguaro hole or "boot." It prefers cholla cacti instead.

Among the forks of the saguaro branches one

is likely to see a hawk's nest. From this lofty vantage point, hawk parents can keep watch over their offspring and at the same time keep an eye on the ground, where prey, perhaps looking for an opportunity to become part of the saguaro community, may show up. The white-winged dove also frequents the saguaro. The dove feeds upon the nectar of the blossom and the fruit of the plant, making it a major pollinator and propagator of this giant of the desert.

Unfortunately, the indigenous birds that depend upon the saguaro for nesting sites are being threatened by an interloper: the starling. Here we have an example of what can happen to the delicate balance of an ecosystem when a non-native species (and therefore not a natural part of that ecosystem) comes upon the scene. The starling was introduced into the eastern United States from Europe in the late nineteenth century. Since then these birds have migrated west and now live in parts of the Sonoran Desert, where they have begun to displace many native birds by taking over woodpecker nests and competing with native Sonoran Desert birds for food.

PETER KRESAN

A Gila woodpecker carries nesting material to its "apartment" in a desert "condominium," a saguaro cactus.

DANIEL L FISCHER

A screech owl peers from its saguaro home.

TAD NICHOLS

Competition to harvest the delicious saguaro fruit is intense and involves many desert dwellers, including people. The sweet fruit has the consistency of a ripe fig.

Adding to the mystique of the saguaro is the curious fact that pollination, which usually occurs in the month of May, must take place within a period of less than twenty-four hours. Each flower opens after sundown and becomes fully unfurled by midnight. It stays open much of the following day and then closes, never to bloom again. But during its brief moment of life, the saguaro blossom has been host to a myriad of creatures that participate in a grand and glorious feast. At dark, night-flying creatures—mainly nectar bats—flit from one flower to another to partake of the bounty; in the daylight hours, ants, wasps, bees, butterflies, and birds come to take their fill. By June the pollination ritual of the saguaro is complete. The fruit then

© ALLAN MORGAN

The amazing kangaroo rat never has to drink water! This rodent (not really a rat) is the epitome, in both behavior and physiology, of the desert-adapted mammal.

appears, triggering another cycle of gluttony, one in which another desert dweller—man—figures prominently.

The Tohono O'otam (Papago) Indians of the area have harvested the delicious fruit of the saguaro for centuries. Using long poles made from the ribs of the saguaro itself, the Indians knock the fruit from the cactus before Sonoran Desert animals have had a chance to get it all. The fruit—a three-inch, greenish, egg-shaped capsule that splits open to reveal a bright-red, sweet pulp imbedded with innumerable tiny black seeds—is used by the Indians to prepare syrup, jam, and ceremonial wine. Their ancient food-preserving techniques are often observed during the harvest season by fascinated onlookers in the workshops that are conducted, with the help of the O'otam, by the Arizona-Sonora Desert Museum.

But it is primarily insects and the winged and four-footed animals that benefit from the saguaro fruit. The day-active Harris antelope squirrel, the rock squirrel, packrat, pocket mouse, and even the cactus mouse (who eats mainly insects) and other nocturnal rodents either flock to the fruit after it falls to the ground or climb up the saguaro to reach the delicacy. Larger mammals such as the coyote, fox, skunk, and javelina are also attracted to the feast, as well as almost all desert birds. Predatory birds do not partake, but even they reap the benefits of the harvest. The hawk that circles over the saguaro and the owl that lives inside the cavity have only to wait patiently; sooner or later their prey will happen by, looking for an easy meal of saguaro fruit!

Among the banqueters is an animal that deserves special attention. The kangaroo rat—neither a kangaroo relative nor a rat—exhibits the most striking of all arid-land adaptations: the ability to survive *without ever having to drink water!* This remarkable achievement is due to the rodent's highly developed kidneys (many times more efficient than human kidneys). Its urine is highly concentrated and the feces are extremely dry because the moisture content of the feces is reabsorbed by the large intestine. But how does the animal get any moisture in the first place? The answer lies partly in its metabolism. Vital fluids are provided through the metabolic breakdown of the plants and seeds upon which the kangaroo rat feeds. Although this process is not unique to this rodent, the unusual characteristic is that the kangaroo rat can survive without any additional fluids. Other factors help this animal to survive without drinking. Like all rodents, it has no sweat glands and thus loses no body water through its skin. And, in its burrow, humidity is maintained by the moisture content of the animal's exhaled breaths.

The kangaroo rat is a veritable arid-land machine—in its lifestyle as well as its anatomy. It (of course) has very long hind legs. It also has a bounding gait that, apart from providing the animal with speed, serves to reduce the area of the body and the time in which the body is exposed to the hot surface of the ground. It uses its efficient forelegs for digging the extensive burrow system in which it will live during the day and where it will store bushels of seed. Most kangaroo rats locate their burrows under clumps of grasses or shrubs. Smart house-building! The roots of the tussock grasses consolidate the soil, thereby reducing the danger of part of the burrow system collapsing. When its young are born in the spring, even this well-adapted desert rodent needs extra water—so that it can produce milk. The kangaroo rat obtains this moisture from the juicy grasses that grow during this spring season, augmenting them with cacti and other succulents.

Although that owl in the saguaro is quick to spot and swoop down upon the kangaroo rat on the ground, the rodent will probably hear him coming and thereby escape. Thus the kangaroo rat exhibits yet another adaptation, one that allows it to detect approaching predators such as birds, reptiles, and mammals. In fact, its keen hearing is four times more effective than that of a human being. It can hear sounds as subtle as the scraping of a snake upon the sand or the flapping of an owl's wings. When it does detect a predator, the kangaroo rat puts those springy hind legs into motion and is gone in a flash. A truly astonishing little beast!

With nearly everything that lives nearby reaping the harvest of the saguaro fruit, it would appear that there is little chance for the plant to reproduce. But the saguaro, as the generous host of so many living things, is too valuable for nature to do without, and so she has found a way to ensure the survival of at least some of the plant's progeny.

The secret lies in the numbers of seeds. Each plant produces approximately 200 fruit per year, and each saguaro fruit contains an average of about 2,200 seeds. This means that, since production of a single plant may continue for a century, a single saguaro may produce forty or so million seeds during its lifetime. Given such odds, it would seem that *many* seeds would survive, even though saguaro seeds are known to be highly prized by such prodigious consumers as the insects of the desert. (Ants alone carry off ninety percent of the seeds before they have had a chance to germinate.) But the number of saguaro plants remains more or less constant, so apparently only *one* of these forty or so million seeds manages to survive to replace its parent.

LYNN M STONE

The coati (or coatimundi), a relative of the racoon, is an endearing resident of the Sonoran Desert region. This tree-climbing mammal spends its days actively searching for anything edible—berries, rodents, insects, eggs.

RICK McINTYRE

The javelina—a peccary, not a pig—is another who partakes of saguaro fruit. This beast also eats prickly pear cacti—thorns and all!

Overleaf: The desert in bloom is splendor unsurpassed. Photo by Jody Forster

Desert Arthropods

The Sonoran Desert claims some of the most intriguing of all insects. They exist in every habitat that the desert offers—from the seashore of the Gulf of California to the summits of the highest mountains. They live in the tops of the tallest trees and at the depths of the deepest caves. They are everywhere.

Among the unique insects of the Sonoran Desert is the velvet ant, a member of the largest group (eighty percent of the total) of animals on earth, the *arthropods.* (Insects, by the way, outnumber all the other arthropods—arachnids, centipedes, millipedes, crustaceans—put together.) Actually, the velvet ant is not an ant at all, but a wasp that lays its eggs within the homes of burrowing bees. Like many wasps, the velvet ant stings, and its venom can cause severe reactions in some people. It's best to keep one's distance!

Another interesting wasp is the tarantula hawk. This insect attacks and paralyzes its favorite target, the tarantula, then drags the victim back to its hole, where the wasp lays a single egg on the tarantula's immobile body. The egg hatches and the larva eats its food supply—none other than the unfortunate tarantula. The larva remains in the tarantula until it pupates and hatches into an adult wasp.

© ALLAN MORGAN

The tarantula, whose image strikes terror in many hearts, is in fact a gentle spider and a valuable asset to the desert community.

The bright-hued longhorn beetle feeds on the oozing sap of desert broom.

STEVE PRCHAL

Then there is the Pinacate beetle, which wards off an attacker by standing on its head and secreting a foul-smelling substance. And tiny insects called çochineal scales, which suck juice from the prickly-pear cactus while protecting themselves with a cottonlike substance they themselves secrete. (Since ancient times, the Indians have used cochineals to produce a reddish-purple dye. Squeeze one betweeen your fingers to see why.) And there are the hairy black flies, which parasitize the caterpillars of some species of moths and butterflies.

Hundreds of species of moths and butterflies inhabit the Sonoran Desert. These beautiful insects help to pollinate and thus ensure continuation from generation to generation of the spectacular plants of the Sonoran Desert. They lack aggressive features, utilizing instead a passive defense, in nearly every known form. Camouflage is the most common defense. It appears to be extremely effective and may involve shape, color, color patterns, and even behavioral traits. Positioning closely against their backgrounds, together with slow, almost indiscernible movement, enables these insects to blend unseen into their surroundings.

Mimicry, another form of passive defense, allows some moths and butterflies to appear to be unappetizing or even dangerous to their enemies. A butterfly that looks like an unpalatable species, for example, will not be bothered by birds. Some not only appear to be unpalatable but *are.* And there is another clever little ploy: the ability of some species to display eyespots whenever their enemies approach. The spots resemble the large eyes of an owl, and since most birds are fearful of that fierce predator, they avoid anything that remotely resembles it.

Some desert insects depend upon the summer rains for their existence. The ephemeral plants produced in the aftermath of rain are the only source of food for many insects. A number of these species spend much of their lives in the form of eggs. When the rains occur and the vegetation appears, they hatch. The three-inch-long, black paloverde root borer, for example, remains in the ground until it is an adult. At the beginning of the rainy season it emerges, seeks a mate, lays its eggs, and dies. As an adult it never eats.

Insects provide a nearly unlimited food supply for other animals, and arachnids—spiders, scorpions, pseudoscorpions—prey heavily upon desert insects. One of the best known and most misunderstood of this group in the Sonoran Desert is the tarantula, a gigantic, hairy spider that averages about two inches in size, and has legs that more than triple its length.

© ALLAN MORGAN

The queen butterfly is only one of hundreds of butterfly species that inhabit the Sonoran Desert region.

The size and singular appearance of the tarantula make it an easy target for the writers and producers of horror movies, who often seize upon this spider as the perfect creature with which to satisfy thrill-seeking audiences. Thus another myth is perpetuated. That is a shame, because the tarantula is an extremely important link in the desert ecological chain, and it is not threatening to humans. One must really go out of his way to provoke a tarantula and get it to bite, and then it will do so only to protect itself. The tarantula, by the way, has fangs and kills its prey with its bite, not its venom, which has little effect, even on human beings.

LYNN M STONE

The collared lizard, one of the largest and most colorful of the desert's lizards, sometimes runs on its hind legs.

Tarantulas are incredibly long-lived. The spiders don't mature until they are about ten or eleven years old, and it is at this time that mating takes place. The myth that the male dies after mating is just that, another myth. True, the male may die after mating, but that is because they do not live beyond the year in which they mature. Male tarantulas, which are slightly smaller than females, live to be about twelve or thirteen years old; females may live twice that long, or about twenty-five years.

For the first few years of its life the tarantula molts several times a year, then annually after that. The tarantula is extremely vulnerable during these periods. If while molting a tarantula breaks itself, it is likely to bleed to death. Tarantulas, like other spiders, replace lost appendages when molting. Most spiders, including the male tarantula, molt only until they reach the adult stage. The female tarantula, however, molts annually well after maturity, replacing appendages and her reproductive system each year. One expert has written that it can be said the female tarantula annually restores her virginity.

The tarantula is not a protected species, nor is man endangering its existence. In recent years a mail-order concern has made a business of specializing in tarantulas as "pets," but that is more bothersome in an exploitative sense than it is threatening to this spider's continuation as a species.

Another fascinating arachnid of the Sonoran Desert is the nocturnal scorpion, an aggressive predator that has *earned* its dangerous reputation. The sting of a bark scorpion can mean serious trouble for an infant or elderly person. The bark scorpion is smaller than the giant hairy scorpion, whose sting is much less serious. The scorpion gives birth to live young. The Arizona-Sonora Desert Museum has an excellent scorpion exhibit, which occasionally includes a mother with newborn babies clinging to her body.

The Sonoran Desert, like many other areas in North America, is host to two of the most dangerously venomous spiders on the continent—the black widow and the brown spider. The black widow female *does* kill the male after mating, so that is at least one story that is true as well as popular.

Also worthy of mention is the centipede, a fast-moving, venomous, segmented brute that averages eight inches in length and possesses up to forty pairs of legs. Not the sort of thing one likes to find in one's living room, but this is not an uncommon occurrence if your house is located in the Sonoran Desert.

© ALLAN MORGAN

Because of its common name many people believe that the venomous giant desert centipede has a hundred legs. Actually, it has less than fifty legs.

Rattlesnakes and Roadrunners

Possibly no creature (unless it is the roadrunner) conjures up images of the desert more than does the rattlesnake. The Sonoran Desert region is home to a number of species ranging in size from the western diamondback, up to six feet in length, to the two-foot-long forms that live in the mountains.

(All rattlesnakes are pit vipers, so called not because they live in pits but because they have a deep hollow, or pit, on each side of their head. The pits are heat-sensing devices, which detect the presence of a possible meal in the neighborhood.)

The most unique of the rattlesnakes is the sidewinder. Averaging about eighteen inches in length, this viper gets its name from the way it travels rapidly and gracefully across the desert in a sideways, looping motion, leaving in the sand the unmistakable traces of its passing.

One might imagine that rattlesnakes have few predators, but this is not actually so. Young snakes, venomous or not, must constantly be on the alert against attacks by other snakes, some mammals, and especially birds of prey.

One of the young snake's most dangerous predators is, in fact, the roadrunner—the picturesque bird the movies have made so familiar and lovable to all of us. True to its reputation, this desert cuckoo is swift of foot. It runs in such a de-

TOM BRAKEFIELD/ANIMALS ANIMALS

The roadrunner, that most captivating of Sonoran Desert residents, frequently can be seen dashing across roads and trails.

termined and streamlined manner (with the tuft at the top of its head in the "down" position) that it gives the impression of having important business to take care of and a destination well in mind. But when the brakes go on, this two-foot dandy of the desert appears to don a bemused expression, as if surprised to be at such a place and wondering how and why it got there!

The roadrunner, who also feeds on lizards and mice, uses its tremendous agility and energy to worry and wear out its prey. Once a young snake is exhausted from striking at the roadrunner, the bird merely moves in and with his long beak helps himself.

Only in cartoons does the roadrunner say "meep, meep," and only in cartoons is he constantly being chased by a not-so-smart coyote.

The western diamondback, largest and most numerous rattlesnake species in the Sonoran Desert, blends unobtrusively with its surroundings.

C. ALLAN MORGAN

This pattern in sand points unmistakably to its originator, a sidewinder rattlesnake, and suggests the undulating, lateral looping motion with which this snake travels in its desert environment.

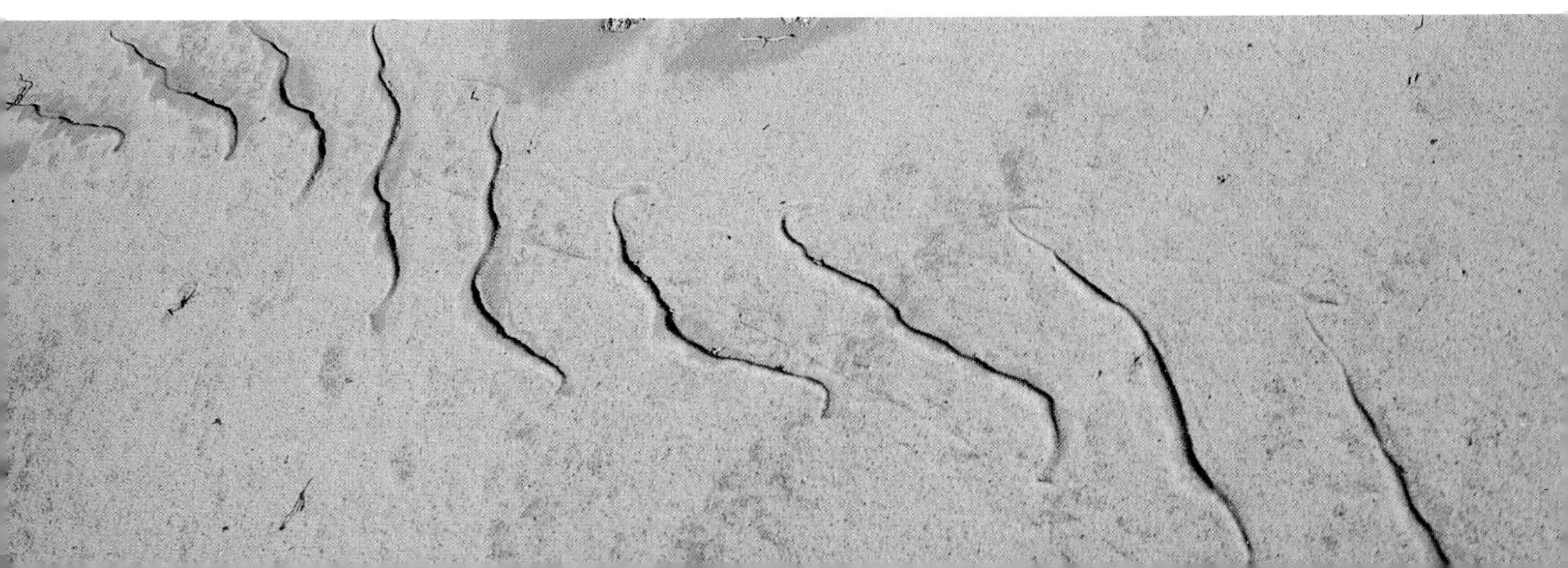

C ALLAN MORGAN

The spectacular jaguar, largest of the North American cats and once relatively common in the American Southwest, now ranges only in the Mexican portion of the Sonoran Desert region.

C ALLAN MORGAN

Scientists are still learning about the behavior in the wild of the secretive margay, agile acrobat of the cat family.

The Cats

Small, medium, and large mammals are prevalent throughout the Sonoran Desert. The numerous ecosystems of the region support black bears, bighorn sheep, beaver, otter, and other mammals one wouldn't expect to find in a desert. Then there are those most fascinating animals of all—the cats.

It's not surprising that few people have ever viewed in its natural habitat *any* representative of the six species of wild cats that live in the Sonoran region. Under cover of darkness and protected by the color of its fur, a cat becomes nearly invisible. If one *is* lucky enough to spot a Sonoran Desert cat in the wild in the United States, it would probably be a mountain lion, bobcat, or jaguarundi. The other Sonoran cats—jaguar, margay, and ocelot—occur in the more remote southern parts of Sonora, Mexico.

Zoologists divide the true cats into two categories according to size—small and large. Differences between the two groups are anatomical as well as behavioral. Large cats have a tongue support that enables them to roar. Some can also purr, but only when exhaling. Small cats cannot roar, but can purr when both inhaling and exhaling. Large cats always feed in a prone position; small cats feed in a crouched position. Watch your housecat. The next time it catches a mouse, notice that it holds the mouse to the ground with its paws while eating; large cats don't do this. Another difference: all small cats frequently and thoroughly groom themselves; large cats indulge in this activity much less often and then do only a cursory job.

At any rate, all Sonoran cats are classed as "small cats," except for the jaguar, a magnificent "large cat" that ranges widely from northern Mexico, in the Sonoran Desert, down to Argentina. The jaguar is a good tree climber but prefers to spend most of its time on the ground, stalking peccary herds and watching for a chance to pick off a straggler. Other prey include deer and large rodents as well as fish and turtles, which the jaguar hooks out of the water with a quick swoop of its powerful claws.

The jaguar is a traveler, and sometimes a male—perhaps driven from home in a territorial dispute with another male—may migrate for hundreds of miles. A male jaguar, killed in 1955 near the southern tip of the San Pedro Mártir range in Baja California, is known to have wandered across the entire Sonoran Desert, crossed the Colorado River, and traveled south for a hundred miles—a total distance of some five hundred miles from regularly occupied jaguar range!

Jaguars once were fairly common in the American Southwest. But hunting, trapping, and the expropriation of much of its natural habitat by man has resulted in the almost total extinction of the jaguar in this area.

Only two of the six species of spotted small cats from South America—the ocelot and the margay—range as far north as the Sonoran Desert. The ocelot, because of its strikingly beautiful fur, has been hunted more extensively than any of the small cats. As a result, it has become extremely rare over large portions of its range. Since 1973 the ocelot—and all spotted cats—has been on the endangered list. Fortunately, government regulations allow the importation of this rare and magnificent animal to qualified institutions, but strictly for purposes of research, breeding, display, and education.

The ocelot's closest relative, the margay, or *tigrillo* as it is known in Mexico, is an elusive forest dweller and probably the most accomplished acrobat of all the cats. A special anatomical adaptation of its joints allows the margay to rotate its hind feet inward as much as 180 degrees. This provides the margay with a remarkable agility. It can hang from a branch by its hind feet or by its wrists; it can move upside down along a branch in sloth fashion; and it can jump off a tree and catch itself, by means of its extended claws, on a branch far below. No other cat can run headfirst down a tree trunk; the margay can do so in either a gallop or diagonal walk.

The jaguarundi, or *leoncillo,* with its elongated body and relatively short legs, looks more like a weasel than a cat. Its fur is short, smooth, and without pattern. Several color phases occur: from gray and reddish animals to brown and blackish individuals. The jaguarundi is occasionally found in the southern part of Arizona and all through Sonora, Mexico, in areas near river bottoms and in semidesert grasslands transected by arroyos. It preys mainly on rodents and birds.

The markings of the ocelot—and the other spotted cats—are both a blessing and a curse for this splendid animal. The coat allows the ocelot to blend unseen into the terrain of its habitat, a definite advantage in stalking prey, but its remarkable beauty makes the cat a prime target for relentless fur hunters. The Endangered Species Act makes the importation of the pelts of all the spotted cats illegal in the United States.

C. ALLAN MORGAN

The bobcat, or *gato del monte,* is a common wild cat that inhabits scrub, thickets, broken country, lowlands, and the desert, but is never found in entirely coverless terrain. The bobcat maintains a well-marked territory with fixed trails, toilet spots, and rest sites; it dens in a rocky crevice, brushpile, or hollow tree trunk.

The ocelot, the margay, the jaguarundi, and the bobcat are exhibited at the Arizona-Sonora Desert Museum in naturalistic habitat settings that have been lauded as some of the best exhibits of their kind in the world.

ASDM PHOTO BY CHUCK HANSON

George L. (for Leo) Mountain Lion helped publicize the Arizona-Sonora Desert Museum when it was founded in 1952. "George" is still the symbol of the museum.

The largest of the small cats is the mountain lion, also known as puma, cougar, catamount, panther, and—at the Arizona-Sonora Desert Museum—as George L. Mountain Lion. It was "George" who, through a regular newspaper column (ghost-written by the museum's co-founder William Carr) helped Tucsonans become more knowledgeable about the Sonoran Desert and the museum. Through the years, this column also boosted the museum's reputation internationally.

The mountain lion occurs from British Columbia, Canada, to Argentina. Its range is the most extensive of any of the cats in the Western Hemisphere. These solitary creatures have a life span of about twenty years—*if* they are lucky enough to avoid man, their most dangerous enemy. Pumas are secretive animals and show themselves infrequently, although some sightings around populated areas have been reported, when for one reason or another a puma has wandered out of the mountains.

A puma on the prowl is a fascinating study in grace and motion. As the animal stalks its prey, it keeps its body low to the ground, moving with great stealth and cunning. Suddenly, the puma pounces, efficiently dispatching its unfortunate victim with a well-aimed bite into the nape of the neck. The mountain lion is an excellent sprinter and can develop considerable speed over short distances, but it tires easily and will give up pursuit if it cannot overcome its prey quickly. This cat is also an excellent jumper and has been known to jump as far as twenty-two vertical feet.

The ringtail is sometimes called "miner's cat" in the Southwest because of its usefulness to early miners and prospectors in controlling rodents.

C. ALLAN MORGAN

Among the mammals of the Sonoran Desert lives the mysterious ringtail cat. Why do we say "among the mammals" and not "among the cats"? Here we go again. The kangaroo rat is not a rat. The velvet ant is not an ant. And—that's right—the ringtail cat is not a cat! This small carnivore is a member of the raccoon family.

Known by many other names as well—bassarisk, cat squirrel, miner's cat, civet cat, coon cat—the ringtail cat is commonly found throughout the American Southwest and is widely distributed throughout western North America, from southern Oregon down into the Mexican states of Veracruz and Oaxaca. It lives in rocky cliffs, canyons, chapparral, and occasionally in riparian habitats. Like many of its desert companions, the ringtail cat has developed the extremely practical ability by which, when feeding on an exclusive diet of rodents, it totally loses its dependence upon water. In the course of evolutionary history, the ringtail has not changed its appearance or way of life since the middle Tertiary, some 13 million years ago. It may justifiably be called a "living fossil"; with its set of carnivorous teeth the ringtail is the most primitive member of the raccoon family.

The Boojum and Other Oddities

Speaking of "living fossils," there is no plant on the Sonoran Desert that *seems* more deserving of this designation than the boojum tree—no doubt because the boojum is such a strange sight that one assumes it is a relic of prehistoric times. It is certainly the most curious of the curious. But, although the boojum may *look* primitive, it is actually a highly advanced form of life and belongs very much to the here and now. This tree grows in only two limited areas—central Baja California and the area south of Puerto Libertad in Sonora, Mexico—both of which are mainly characterized by barren hills and alkaline valleys.

The boojum has often been described as an "upside-down carrot," obviously a gigantic one, since it reaches a height of sixty feet or more. These grotesque but strangely beautiful giants dwarf all their non-cactus associates. It is no wonder that people who are seeing the boojum for the first time are struck speechless—at least for the moment. There is simply nothing that compares with the boojum, unless it is an inhabitant of the world of Dr. Seuss. Branches sprout from the top of the tree—sometimes growing straight, sometimes perpendicular, and occasionally twisting around another branch. In some cases the branches grow so long that they bow and arch toward the ground, like giant catapults ready to spring into action. Smaller, thorny, twiglike boojum branches project outward from the tree's stem, becoming green-leafed only when water is available for the roots to suck up, a plant adaptation known as "drought dormancy." In effect, drought-dormancy plants such as the boojum shut down their metabolism whenever water is scarce, preventing loss of moisture.

PETER KRESAN

In 1922, during an expedition organized by the Desert Botanical Laboratory of Tucson, geographer Godfrey Sykes stared at an odd plant near Puerto Libertad, Sonora. "Ho, ho! A boojum, definitely a boojum!" said the amazed Sykes. And that has been the colorfully uncommon common name of this plant ever since. The thornlike twigs that project from the stem of the boojum produce leaves only when there is sufficient water to nourish them.

JEFF GNASS

DAVID MUENCH

Sunlight illumines the forms of that most graceful of plants, the ocotillo.

JODY FORSTER

Springtime blossoms of the paloverde transform the trees into a riot of yellow color.

The *ocotillo,* another peculiar Sonoran plant and relative of the boojum, is an especially striking example of drought dormancy. This apparently lifeless bundle of thorny sticks can grow a full crop of leaves within only forty-eight hours after a warm rain. The leaves themselves are soft and tender; possessing no adaptations for desert life, they fall off the plant once moisture is no longer available. In the warmest regions of the Sonoran Desert, the ocotillo may produce leaves in any month and may go through several leafing cycles in a year.

The paloverde carries this strategy a step further. Its branches are green because the bark contains chlorophyll, the remarkable chemical that captures solar energy and processes it into stored chemical energy (sugar and starch). Stomates in the branches of the paloverde allow the tree to carry out photosynthesis even when it is leafless, at least to some degree. The ocotillo, boojum, and other desert plants also have this ability, but the chlorophyll in their stems is less obvious.

The extent to which some Sonoran Desert plants carry the drought-dormancy adaptation is almost unbelievable. During severe droughts lasting several years, the tenacious creosotes and paloverdes may drop entire branches, one by one, using the saved water and food to keep the rest of the plant alive. They can die down to ground level and yet still recover whenever the rains finally do arrive.

The precious desert rains revive not only the plants but also all other living things. After a passing thunderstorm, fairy shrimp and tadpole shrimp suddenly appear in the temporary rain pools along the waterways in the desert. The eggs of these small crustaceans remain dormant sometimes for years, surviving the dryness and heat of the sun and hatching only when the rains come.

The spadefoot toad is just as patient. Buried in the soil, it also passes through drought and the cooler months. How does this toad know when it rains? It *hears* the drops patter on the ground surface above! And when this happens, the toad digs itself out and immediately begins a noisy nighttime orgy. Within only two short weeks—before the pond has completely dried up in the summer heat—the spadefoot-toad eggs have hatched and the tadpoles have metamorphosed into land-dwelling toads mature enough to dig their way back into the sand when the rainy season ceases.

Lifeblood of the Desert

Rain gives life to the desert. It miraculously transforms a waiting, patient ecosystem into a world in which activity abounds in every niche. All living things respond in a feverish campaign to take advantage of every moment in which they can replenish nutritional supplies and thus renew their often tenuous hold on life. Every link of the food chain is affected. Soil soaks up the nutrients; plants gorge themselves; insects feed upon the plants; amphibians eat the insects; birds, mammals, and reptiles prey upon the amphibians and each other. In this process some living things must perish, but that is nature's way of revitalizing all the species. And so they live on—month to month, season to season, century to century, millennium to millennium.

In no other region of the Sonoran Desert is the effect of water demonstrated more dramatically than in the Lower Colorado Valley—the hottest, driest, and largest region in the desert. (Ironically, it is this region that produces the most spectacular of any wildflower displays seen in the Sonoran Desert.) Most of this area is sandy or gravelly plains; the rest is highly diverse country—low mountains of varying rock types, sand dunes, and alkali sinks. Washes (dry drainage channels) are prominent in the Lower Colorado Valley because they contrast so greatly with the barrenness of the surrounding flats. The washes here drain extensive areas, and even though they carry runoff only once or twice a year, they store enough water in their deep gravel beds to support the impressive ribbons of trees and large shrubs that follow their braided channels. Thus, there is more vegetation in the washes than elsewhere in the desert, and this difference is reflected in the wildlife. Birds, for example, are about ten times more prevalent in the washes than in the creosote flats.

PAUL R. JOHNSON

A lovely but bare ocotillo is the sole witness to a brilliant desert rainbow that promises life-giving rain. Within days of such a rain, the ocotillo produces leaves.

Many plants cope with unfavorable condi-

PETER KRESAN

The fishhook barrel cactus, whose Latin name appropriately means "fierce," produces blooms that decorate the plant with glorious bursts of fiery color.

Against a craggy backdrop—the Ajo Mountains near Organ Pipe Cactus National Monument—rain falling on arid land has produced a stunning green carpet interwoven with the gold of Mexican poppies. The bloom adorns the desert for only a few weeks and then is gone. It will take another rain at exactly the right time to produce another of these beautiful desert tableaus.

tions simply by *avoiding* them. These are the annuals, plants that complete their cycles in one growing season. Desert annuals live a much shorter time than those growing in other areas. Winter annuals last four to six months; summer annuals are ephemeral, often lasting as little as one month.

Annuals require open, sunny sites free from competition with perennials; annuals are thus more numerous in arid lands than in any other habitat. The percentage of annual species in a regional plant list ranges from zero in the wet tropics to over ninety percent in the driest low deserts. The flats near Yuma, Arizona, have one or two species of perennials and ten to twenty annuals.

The weather-resistant seeds of the annuals may lie dormant in the soil for decades, awaiting suitable growing conditions. They will not germinate until survival to maturity is a certainty. The plants appear in large numbers only in rare wet years; in dry years there are none. These facts are amazing enough, but even more incredible is the fact that each annual seed germinates only when it is the *right* season for it to do so. Winter annuals don't respond to summer or even late-winter rains, and summer annuals do not respond to winter rains. Each species responds only to the particular combination of rain and temperature in which nature intended it to grow.

In a given year and location, the species of annuals predominating in an area can vary over distances of only a hundred yards or less. These differences are influenced by factors such as soil type and texture, slope, and exposure. Some species grow only in sand dunes; others grow only on rock slopes or on limestone. Some grow in the open; others prefer the shelter, however slight, of a shrub or tree. Also, different species may appear in the same spot in different years. There are so many factors involved that species locations cannot be predicted even from year to year.

In the Lower Colorado Valley and in the Arizona Upland, a good bloom occurs only once or twice in a decade. And when the conditions are right for the bloom, there is no more beautiful a land to be found anywhere. Flowers literally carpet ground that was bare only a short time before. The glorious scene is all too brief, however; within three or four weeks the flowers are all gone, and once again only the seemingly barren soil can be seen.

But while the spectacle lasts, the astonishing variety and abundance of the desert wildflowers is overwhelming. Yellow Mexican gold poppies,

The Schott yucca shields its delicate blossoms with an encircling fan of extremely sharp leaves. The yucca—a lily, not a cactus—depends for pollination upon moths, a different species of moth for almost every species of yucca.

The organpipe cactus grows only in a small area of the United States—southern Arizona near the Mexican border. It is prolific, however, in the Mexican portion of the Sonoran Desert. Like the blossoms of the saguaro, organpipe blossoms open only after sundown.

blue lupine, and purple owlclover cover the landscape. Many of the common names of the flowers here are as quaint as their blooms are beautiful: fairy duster, bloodweed, peppergrass, clammy weed, silver puffs. Even locoweed, the plant that causes some livestock to "go loco," produces a beautiful blue or purple flower. The wildflower bloom is the Sonoran Desert's moment of glory, an opportunity to preen itself. And so it does—in lavish abandonment.

The wildflower tableau acts as a lure that draws us to the desert. People come from hundreds of miles away to photograph, paint, or just to enjoy the magnificent richness of the scene. But once here, we become aware of other aspects of this land that we failed to notice before. For there is an inherent beauty here that is as mysterious and fascinating as any tropical jungle. And it will beckon to us again and again over the years.

Silver puffs are spring ephemerals whose name describes their seedheads, not their blossoms, which are yellow.

TAD NICHOLS

The Baja fairy duster responds to a benevolent desert rain with brilliant, tender blooms.

PETER KRESAN

The beauteous ajo lily has a rather unimaginative name. It is, however, an appropriate one: Ajo *is the Spanish word for* garlic, *which also is a member of the lily family.*

Timely spring rains produce a kaleidoscope of striking color: gold poppies, owlclover, desert dandelions, and goldfields. To see the show, people flock to the desert, which gives an inspired and passionate performance that is over all too soon.

PETER KRESAN

JOSEF MUENCH

A profusion of cardón cacti forests the desert near Guaymas, Sonora, Mexico. Found in the southern part of the Sonoran Desert region, these giant cacti dwarf everything in sight, even the majestic saguaro.

TAD NICHOLS

A blooming cardón cactus and the cool sea present a seemingly incongruous juxtaposition. The cardón fruit is highly prized by the Seri Indians of Mexico.

Shore and Sea Life

One of the most difficult facts for strangers to the Sonoran Desert to comprehend is the reality that this *desert* includes an ecosystem that is made up of water, shore birds, and sea life! But truth, as we have already learned, is sometimes stranger than fiction, and on the shores along Sonora and the Baja Peninsula of Mexico and on the offshore islands in the Gulf of California, life exists that is entirely different from that of any other area in the Sonoran region.

True, many of the plants that grow here—ironwood, paloverde, ocotillo, and mesquite—also live in other areas of the Sonoran Desert, but there are some that live here almost exclusively: the boojum, the elephant tree, the cardón cactus (the considerably larger Mexican cousin of the saguaro), the beautiful *paloblanco* (white stick), and the graceful organpipe, a cactus whose fruit is so delicious and nourishing that the natives of the area have been harvesting it for centuries. And in a tiny section of the Magdalena plain, there is a cactus that exists by creeping along the ground in horizontal fashion. From above it looks something like a "miniature" saguaro that has fallen over or a cardón (minus the trunk) whose

arms are reaching for a new body. This strange plant, known as the "creeping devil," roots at the front as it dies back from the rear, relentlessly inching toward some unknown destination.

The islands in the Gulf of California serve as nesting grounds for hundreds of thousands of birds. Isla Rasa, a birdwatcher's paradise, is covered with nesting terns and Heermann's gulls from February through April. Some 200,000 terns and 100,000 gulls have been counted during one nesting season.

This may seem to be an incredible number of birds for only one island, but it actually indicates a deteriorating situation. Less than fifty years ago, the birds of Rasa exceeded one million! Man, as usual, is the villain. Modern transportation, enabling egg hunters to get to Rasa more easily than in the past, had by the early 1960s caused exploitation to reach such proportions that the terns and Heermann's gulls were faced with extinction.

The fate of these birds took a turn for the better, however, when the Arizona-Sonora Desert Museum and the California Academy of Sciences stepped in. Through information they disseminated, the public became aware of this alarming situation, and interested people began to bring pressure to save the birds. The Mexican government investigated the problem and responded swiftly and effectively. In 1964 Isla Rasa was declared a bird sanctuary. Since that time the museum has continued to work closely with the Mexican authorities in providing expertise and money to aid in the efforts on Rasa — efforts that must be carried on continuously if poachers are to be kept away from the tern and gull eggs.

This work was especially important because it inspired a movement to set aside *all* of the islands in the Gulf of California as wildlife refuges. In this movement, as it did with Isla Rasa, the Arizona-Sonora Desert Museum, with the California Academy of Sciences, acted as initiator and leader, although many other individuals and institutions such as the National Audubon Society also participated. Their efforts were rewarded when, in 1978, President Jose López-Portillo of Mexico issued a decree that set all the islands aside as wildlife refuges.

During the nesting season, these islands are anything but peaceful. The birds compete aggressively for space. Isla Rasa, for example, contains only 250 acres and rises only about a hundred feet above the water—not much room for the thousands of birds that fight for nesting privileges there. This is not just a game. It is deadly combat, and it takes place on Rasa every year.

The war goes something like this. Gulls arrive first and lay their eggs. Before long, the terns arrive. Determined to repossess their nesting spots, they drive the gulls out during the night. The gulls retreat but then counterattack in large numbers, recapturing some nesting sites and gorging themselves on tern eggs, which have been unattended in the confusion. The battle is repeated night after night and day after day. It

TAD NICHOLS

The blue-footed booby (left) and the brown booby (right) share Isla de San Pedro Mártir, *one of the Gulf of California islands that the Mexican government set aside in 1978 as a wildlife refuge. These exotic birds got their unflattering name from sailors who thought they were incredibly stupid to land on ships and allow themselves to be caught.*

DANIEL L. FISCHER

DANIEL L FISCHER

A Heermann's gull soars in the air above Isla Rasa. *Gulls are scavengers. Using their strong hooked beaks, they pick up nearly anything they can spot floating in the water.*

finally gets worked out, but not before the number of eggs of both species, gulls and terns, have been considerably reduced. And, although nesting resumes, the animosity continues. Occasional skirmishes—hit-and-run attacks—take place even after both species have hatched their eggs. Gulls prey heavily on tern hatchlings when the adults—off to fetch fish for their offspring—leave the nests unguarded. Man is not the only species who engages in vicious territorial battles!

Other graceful, glamorous, and sometimes strange sea birds contribute to the wonderful menagerie that is present on and around the islands. Boobies, gannets, tropicbirds, brown pelicans, and cormorants nest on Isla de San Pedro Mártir. Fish-eating ospreys hover above the sea, beating their wings slowly and deeply in their search for prey near the surface. Peregrine falcons sometimes soar the skies along with them.

Many of the sea turtles of the Pacific, off the coast of Mexico, are now in extreme danger. The story of their decimation is a tragedy. The coast once provided food, protection, and nesting beaches for one of the largest and most diverse assemblages of these sea turtles in the world. The Seri Indians of the Sonora coast hunted them for centuries, but their use was light. In recent years, however, egg poachers and turtle hunters—exploiting the reptiles for their meat, oil, and shells—have nearly finished them off. Of the millions of sea turtles that once populated the water and coast of this region, only a few hundred thousand remain. It is hoped that ongoing research and stricter laws governing poachers will begin to reverse this lamentable situation.

Complicating the issue are the demands of

California sea lions—actually seals with ears on the outside of their heads—come to the Gulf of California islands to breed and to give birth. The sea lions have little competition here. Their major predator is man, but on these islands they are protected.

DANIEL L FISCHER

Plowing a broad furrow in the wet sand, a green sea turtle lumbers back to the sea after laying its eggs. Only one to two percent of the hatchlings will reach adulthood.

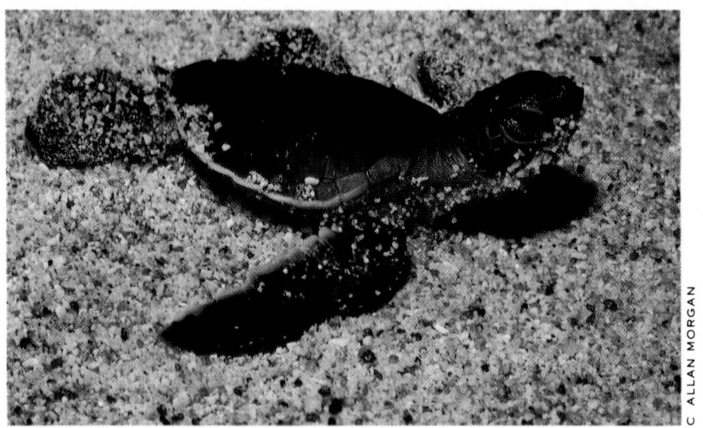

Immediately after hatching, a baby green sea turtle, speckled with wet sand, heads for the sea.

commercial interests, demands that are supported by the natives who earn a living from the turtles. If ways to produce satisfactory substitutions for turtle products can be found, and if the Mexican government can win the support of the Mexican public in order to effect legislation, the turtles may be saved. The turtles, after all, should be the most important consideration. At this point, their future looks bleak.

A more cheerful story is that of the California gray whale. The whales today provide one of the most incredible natural-history spectacles imaginable, as they arrive in early spring to breed, calve, and nurse in the protection of the coastal lagoons off the western coast of Baja California. This was not always so. In the 1930s the gray whale was declared nearly extinct, and a ban was placed on hunting. Fortunately, this occurred early enough to do some good. There are now some 13,000 of these gigantic mammals in existence, a real success story for the conservationists—not to mention the whales.

The future of the gray whale's cousin, the sperm whale, is less certain. Indeed, this mammal seems to be slipping rapidly toward the depths of oblivion. That is, unless researchers can come up with a product that will replace the sperm whale's oil, highly coveted in the marketplace. The best hope for such a substitution appears to lie in a plant that grows throughout the Sonoran Desert. The female of the *jojoba*, a woody evergreen shrub, produces an acornlike nut that contains a liquid wax similar to the oil of the sperm whale. The natural life span of the jojoba is very long—possibly as much as two centuries. Such longevity could mean a lot of oil from just one plant. Research is now being conducted to determine whether the plant can be grown commercially and thereby produce enough oil to meet the demands for the oil now being taken from the sperm whale—which is, bear in mind, a *dead* whale.

Bizarre plants and exotic sea creatures are not the only strange species that exist in Baja California. In 1952 a team of biologists found a rattlesnake on the uninhabited island of Santa Catalina. This was a very unusual rattlesnake; it had no rattles. At first, some experts thought that the reptile had simply been injured by a predator, but validity of the new discovery was confirmed in the mid-1960s when several other specimens were found on this island. Thus, Santa Catalina in the Gulf of California is the only place in the world known to have an entirely rattleless rattlesnake. It is surmised that *Crotalus catalinensis* has lost its rattles because, having no predators, it needs no protection. It lives essentially undisturbed on an isolated island in one of the most mysterious regions in the world—the Sonoran Desert.

An enclosure outside the entrance to the Arizona-Sonora Desert Museum displays several species of native lizards to prepare visitors for a rare experience among thousands of living plants and animals of the desert region.

The Arizona-Sonora Desert Museum

We have written about many things that make the Sonoran Desert unique. And in keeping with its extraordinary characteristics, this desert hosts a singular institution to tell its story: the Arizona-Sonora Desert Museum. Founded solely for the purpose of interpreting the natural history of the Sonoran Desert region, the museum has gained international prominence for the beauty, taste, and authenticity of its exhibits, some of which are being copied throughout the world.

In an article about the top ten zoos in the world, the *New York Times* featured the Arizona-Sonora Desert Museum and described it as ". . . the most distinctive zoo in the United States." The British Broadcasting Corporation produced a series called "The World's Greatest Zoos," which included a half-hour presentation about the Desert Museum. As much a botanical garden and geological interpretive center as it is a zoological park, the museum is the world's foremost regional natural-history institution.

The museum was conceived in response to requests from a number of people who thought that the Sonoran Desert, as one of the most amazing and interesting regions on earth, deserved to be interpreted to the public. They felt that people are entitled to know the desert as it *is*, not what it seems to be, and not what myths have made of it.

The Desert Museum came into being in 1952. It was the brainchild of two forward-thinking men: Arthur N. Pack, who had gained prominence as a conservationist through his work as editor of *Nature Magazine* and through various other conservation groups he had founded, and William H. Carr, who had pioneered in outdoor interpretation in the 1920s while with the American Museum of Natural History.

It was Carr who defined the aim of the museum, envisioning it as a method of employing education "as a means of helping man to recognize and assume his responsibilities toward nature in order to gain some hope of assuring his future." He and cofounder Pack, realizing that education of the public is the necessary first step toward thoughtful conservation, swung into action. The urgency they felt is expressed in Carr's words. "The time for widespread implementation of this kind of endeavor is *now;* before man succeeds in totally defiling his habitat and making it unlivable."

TIM FULLER

Black bears romp in their spacious enclosure which is part of the museum's one-acre Mountain Habitat exhibit. This area introduces visitors to plants and animals which live in habitats 4,500 to 7,000 feet in elevation on "mountain islands" in the Sonoran Desert region. Over 6,000 individual plants representing 100 species are part of this amazing project.

Arthur Pack died in 1975. Bill Carr, the museum's first director, died in 1985. In the short history of the museum, other directors—William H. Woodin, Mervin W. Larson, Holt Bodinson, Dan Davis, and David Morgan Hancocks—have carried on the traditions the founders established.

The Arizona-Sonora Desert Museum is now a vital part of the desert scene. Located approximately fourteen miles west of Tucson, Arizona, it is especially dear to the hearts of the citizens of southern Arizona. It is a private, nonprofit, natural-history educational institution whose purpose continues to be simple and straightforward: to tell the story of the plants, animals, and land of the Sonoran Desert region with the objective of instilling a conservation ethic in visitors who come from around the world to experience its wonders.

Over 300 species of plants are here in natural splendor, and 200 species of live animals are on exhibit in enclosures unmatched for the manner in which they recreate the animals' natural habitats. In February 1986 the Desert Museum opened a mountain-habitats exhibit for mountain lions, black bears, white-tailed deer, Mexican wolves, and several species of birds. Like several of the institution's exhibits throughout the years, the mountain habitats are unique and will no doubt serve as models for similar enclosures worldwide.

Another exhibit that merits attention is the museum's Earth Sciences Center. Visitors enter the world of Sonoran Desert geology through a recreated limestone cave so realistic that many of them argue that it cannot be artificial. They then venture into an area that presents the history of the earth from its beginning 4.5 billion years ago to the formation of the Sonoran Desert region and gives an educated guess as to what the area will look like 50 million years in the future. The story is told with live plants and animals, a necessity from the museum's point of view because the interrelationships of plants, animals, and geology are keys to the understanding of all ecosystems. The minerals of the region—the most beautiful minerals anywhere on earth—also have a special place in the Earth Sciences Center.

GILL C. KENNY

River otters, once common in southeastern Arizona, have been pushed out of their natural habitats. They are a visitor-favorite at the museum.

GILL C. KENNY

Northern Spiny-tailed iguanas roam free on the museum grounds. The newly hatched young are bright green and blend with vegetation during hatching season.

The museum's activities are also important, and they are indeed impressive. In 1990 some 600,000 people visited, including nearly 30,000 Arizona schoolchildren who were there as part of the museum's extensive educational outreach programs in the schools. Other community programs take staff and many of the museum's 160 volunteer docents into hospitals and nursing homes and before service clubs to reach those who are unable to visit on their own. Museum staff and docents also conduct regularly scheduled interpretive talks on the premises every day of the year. In addition, the museum's nearly 18,000 members participate in special-event activities on the grounds and throughout the Sonoran Desert area. Children have the opportunity of attending summer school classes conducted from preschool through junior high.

The Desert Museum does not limit its activities to Arizona. Members can be found all over the United States, and the work the museum does throughout the entire Southwest is noteworthy. In 1974 the museum inaugurated its "Mexico programs." Together with educators and members of public and private sectors in Mexico and the U.S., the museum initiated an across-the-border exchange that resulted in the development of a comprehensive environmental education program in the primary schools of the state of Sonora, Mexico. The museum continues to work closely with Mexico in cooperation with El Centro Ecologico, Hermosillo, a new "sister" institution, to develop mutually beneficial activities and programs.

Most visitors, especially those living in the area, think of the museum as their own, and well they might. Even though the museum is not governmentally funded, the essence of its value is the participation of the public as a whole. The pride of ownership felt throughout the area is apparent in the quality of the museum's exhibits. This pride has been transmitted to and is equally shared in by the museum's dedicated staff members, whose work is a delightful manifestation of their sincerity and enthusiasm.

More than just a pleasant place to spend an afternoon, more than just a learning experience, more than an effective myth dispeller, the Arizona-Sonora Desert Museum is a symbol of the compelling urge to reach out for knowledge, the need to understand our environment, and our desire to have some conception of our own place in the natural world. This is a lofty position to maintain. But the museum does it well, in a complex setting that is harsh yet nurturing, monotonous yet diverse; a desolate yet beautiful region that dominates the great American Southwest—the Sonoran Desert.

TIM FULLER

300 million years of geologic history are interpreted in the museum's Earth Science Center. Limestone caves throughout the Sonoran Desert region contain remnants of ancient life forms which inhabited them.

Golden cholla bloom in the Anza-Borrego Desert State Park of California.

Not to have known—as most men have not—either the mountain or the desert is not to have known one's self.

—Joseph Wood Krutch, *The Desert Year*

Back covers: The Sonoran Desert exhibits a beauty that is uniquely its own. Photos by Jeff Gnass and David Muench

For a list of other Story Behind the Scenery publications write: KC Publications, Box 14447, Las Vegas, Nevada, 89114.

Printed by Dong-A Printing and Publishing, Seoul, Korea